The Magic of Dolphins

The Magic of Dolphins

by

Horace Dobbs

LUTTERWORTH PRESS
CAMBRIDGE

To Matthew & Alison
Julia & Adrian

Lutterworth Press
P.O. Box 60
Cambridge CB1 2NT

British Library Cataloguing in Publication Data
Dobbs, Horace E (Horace Edward),
  The magic of dolphins. – New ed.
  1. Dolphins
  I. Title
  599.5'3

ISBN 07188 2603-5
Copyright © 1984 and 1990 Horace Dobbs

First published in 1984
Reprinted with new material, 1990
Reprinted 1992

Printed in Hong Kong by Colorcraft Ltd.

Contents

Dolphin

Pushing through green waters
Symbol of joy
You leap from the depths
To touch the sky
Scattering spray
Like handfuls of jewels.

Not caged by union rules
Unfettered by sales targets
No trains, or planes to catch
Your time is set by the flow
Of the sea's tide
And the moon's glow.

You give us images of ecstacy
That we lock away
Behind the doors of memory
For quiet moments
When released by our possessions
We dream of freedom like yours.

Horace Dobbs
Portreath July 1982.

Prelude

A boy on a dolphin

IT WAS late afternoon. The sea around us was pale green and I could hear the bubbles from my aqualung gurgling towards the surface. Out of the haze a grey shape appeared. It hurtled past me and disappeared in a trice. I looked up. The sun, seen through the ruffled silver surface of the sea, looked like a deep orange disc. In the middle of the disc was the black silhouette of my thirteen-year-old son Ashley. He was wearing a wetsuit made of sponge rubber and wore a facemask that enabled him to see clearly into the water. He turned his head from side to side to see where the dolphin would appear next.

Still looking up, I sat on a rock on the seabed and watched Ashley move slowly across the surface about six metres above me. Suddenly, I was aware of a presence. A few seconds later a plume of bubbles exploded into the water beside Ashley. Out of them I saw the grey torpedo shape of the dolphin appear; then, with a few powerful up and down thrusts with his tail, he disappeared like a greyhound after a rabbit. Moments later the dolphin rushed past me and hurtled towards the surface to launch himself once again high into the air.

I was conscious of the enormous power that was unleashed as the dolphin leapt over Ashley, and I became very concerned for the safety of my son. What would happen if the dolphin misjudged his jump and landed on the boy? Ashley could be injured or killed.

Pushing my large camera housing in front of me, I rushed for the surface to tell Ashley to get out of the water and climb into our small inflatable boat. As I broke the surface I saw a sight which has remained in my memory as if it were carved in stone. I saw Ashley rise slowly out of the water with his legs

Ashley Dobbs and Donald the wild dolphin off the Isle of Man.

astride the head of the dolphin. At first the boy looked slightly scared and incredulous. Then he broke into a grin and held both arms aloft – like a footballer who has just scored a goal. As he did so, he moved away from me.

Ashley was perfectly balanced on the dolphin's head. The dolphin, with Ashley riding like a jockey, circled the harbour and came back alongside me, before suddenly diving to leave the boy once again floating on the surface. The sight of my own son riding a wild dolphin in the sea is one of my most cherished memories. It brought the dolphin, with whom I had developed a close friendship, and my son who I loved dearly, into harmony with one another and the sea around us.

It was a moment which revealed to me one more of the many facets of the mysterious magic of dolphins.

The full story of my many experiences with Donald the dolphin are told in my book *Follow a Wild Dolphin*, and its sequel *Save the Dolphins*. In this book I am going to look at what we know about dolphins, and identify the qualities that give them a very special place in our affections.

1

The Nature of Dolphins

THE IDEA that there is something special about the dolphin is not new. The Minoans knew about it as far back as 1500 B.C. To them dolphins were symbols of joy and music. Centuries later the ancient Greeks dedicated their temple at Delphi to a dolphin god (dolphin comes from the Greek word *delphi*, which means 'womb').

The worship of dolphins would run contrary to most modern religious beliefs. However, there is an element of mystery in all religions. Most of them advocate peaceful co-existence between man and his fellows and the other creatures that inhabit the earth. Although the people of ancient civilizations did not have the benefit of the scientific knowledge we have gained in recent times, they did have a deep love and respect for dolphins. And the remains of their architecture bears witness to the fact that they were in much closer harmony with nature than those who live in cities today.

The parables in the Bible and many of the myths were devised to explain mysteries to those who did not have the knowledge or insight to understand. Like the man inside the dragon's costume in a Chinese pageant, the truth was wrapped in an elaborate story.

So what are the qualities in dolphins that caused them to be so highly regarded by the ancients and still endear them to many people today?

When searching for the answer to that question it is almost essential to think in human terms. By asking ourselves, "What is it I like and admire in people?" and "Do the same type of features exist in dolphins?" we can come towards the solution. Taking an overall view, it soon becomes apparent that dolphins have many of the parts of the human character that we like to find in

A dolphin can deal a death blow to a threatening shark. Divers are unlikely to see sharks when in the company of a dolphin.

our friends; but they seem to lack the traits that we find objectionable in people we strongly dislike.

Lack of aggression must come first and foremost on the list of the dolphin's good points. I can find no incident on record where a dolphin has made an *unprovoked* attack on a man. That is not to say that dolphins are not able to look after themselves. Indeed, sharks usually keep well clear of them because a dolphin can deliver a death-blow to a shark. The dolphin simply swims very fast at the shark and rams it in the side. The blow from the dolphin's beak ruptures the internal organs, and the stunned and fatally wounded shark spirals down into the depths.

Thus the dolphin is well aware that a blow from its beak is lethal. And if it can kill a shark, a dolphin could certainly kill a man in the water with equal ease. Hundreds of people are savaged and maimed by domestic animals every year. We all know how vicious a dog can be if we make it angry. Although face to face encounters with dolphins are obviously much rarer, it is a remarkable fact that they do not attack people.

The special relationship dolphins have with humans seems to extend beyond not wishing to harm them. For there are many stories which tell of wild dolphins helping people in distress – often saving their lives.

But being friendly and helpful towards man is only part of the dolphin's make-up. Dolphins also have a sense of fun. They love to play, as do other wild animals. But most animals limit their play to that part of their life when they are young. This is regarded as part of the essential learning process that will equip them for the serious business of survival in later life. Thus a lion cub will frolic like a spring lamb, but when it becomes mature and takes a senior role in the pride, it becomes much more staid. In contrast dolphins appear to enjoy leaping out of the water, simply for fun, throughout their lives.

However, there is one characteristic which I thought was unique to humans until I met Donald the dolphin, and that was a sense of humour. There were times when it seemed he understood exactly what we were doing. And then he would set out to play tricks on us!

Coupled with the fun and sense of humour of dolphins is another quality: happiness. It seems to come bubbling out of them and infects those who are lucky enough to enjoy their company. Many a sailor has felt heart-warmed when a school of dolphins has joined his ship. Even the most hardened of old sea dogs will break into a smile when recalling dolphins who have escorted him on his voyages. To the ancient sailors the dolphins were the bringers of good fortune. Today, when many people are sailing alone, the sight of dolphins at sea brings joy to their hearts.

I remember the first time I saw a school of dolphins, when sailing amongst the Galapagos Islands in the Pacific Ocean. The cry "dolphins" went up and everybody rushed on deck. A school of dolphins was heading towards us. Leaping joyfully out of the water as they went along, they soon caught up with us. Once they reached our boat the dolphins weaved effortlessly from side to side in the bow wave. They seemed to cast a spell on every-

Dolphins appear to have a sense of humour and spend much of their time playing.

Stylized dolphins have been used to decorate many man-made objects such as this Victorian salt-glaze jug which has a boy-on-a-dolphin motif.

body onboard. We were all captivated. Part of the magic of dolphins is undoubtedly the feeling of joy they impart to people who encounter them – especially when the dolphins are in a playful mood.

Natural beauty takes many forms. A seagull gliding effortlessly in an updraught from a cliff face describes patterns in the sky which are beautiful to watch. And the movement of a school of dolphins rising, glistening, from the water has the same kind of quality. It is beauty based upon mastery and harmony. As the seagull is master of movement through the air, so the dolphin is master of movement through water. That mastery comes through being the right shape and being able to make the maximum progress with the minimum of effort.

14

Imagine being cast ashore without tools and without clothes on land that had never been inhabited by people. Although it might at first seem romantic, most of us would find it very hard to survive, for we have become increasingly dependent upon mechanical aids. Without them, feeding ourselves and keeping warm enough to stay alive would be difficult. The fact is that we have moved out of harmony with nature. We are constantly manipulating the world around us to suit ourselves – often at the expense of the other life forms on the planet. It is not surprising then, when we look at dolphins, that we admire the way in which they blend so naturally with their environment.

One of the qualities needed for the survival of a large animal in the harsh environment of the sea is strength. And the dolphin has strength in good measure. Underwater the dolphin is virtually weightless, so an elaborate bone and muscle structure is not needed to support the body weight as it would if the animal lived on land. This being so most of the dolphin's considerable muscle power is deployed moving it through the water. And nobody would deny that the dolphin is impressive when it comes to physical prowess. Its ability to travel very fast and perform spectacular leaps wins the admiration of all those who appreciate athletic activities, such as gymnastics.

For many people it is the combination of power, grace and beauty that makes the magic of the dolphin. For others it is the dolphin's exuberance and cheerfulness. But there is another ingredient in the mixture which is something more than all of these things. Those who have come very close to dolphins feel it inside themselves, yet cannot explain it. Exactly what it is remains a mystery. For want of a better word let us call it the **spirit** of the dolphin.

Nature still has many secrets locked away and mysteries for us to wonder at. And the spirit of the dolphin is one of them.

2

Evolution of Dolphins

DOLPHINS roam all the seas of the world and some even live in rivers. How long they have been on earth and how they arrived in the first place are the next questions we shall examine.

To do this we must, of course, consider evolution.

Just as a set of plans are required when a new building is built so, too, is a set of plans required when a new body is built, be it human or dolphin. In every living being the plans are laid down in very long and complex molecules called chromosomes. Individual units in the chromosomes are called genes. And it is your genetic make-up that decides whether you will be tall or short, blue-eyed or brown-eyed, have fair hair or raven hair.

In mammals the master plan is created at the moment the egg is fertilized, and it carries a mixture of genes from the two parents. You only have to look at the different faces and shapes and sizes of a class of children of the same age to see how many variations can be produced by this process. There are times, however, when this law is not followed exactly and the offspring is born with some difference not possessed by the parents or their forebears. If that happens, and the offspring is better able to survive as a result, then it is likely that the new feature will be passed on to the next generation. This process is called mutation, and it is through this that evolutionary changes take place.

Most mutants, as the new varieties are called, are less adapted to their environment than their parents. They do not mate and their own particular form is not perpetuated. There are just a few however which can mate and generate more of their own kind.

It is interesting to speculate on the changes that took place millions of years

16

ago and led to the evolution of the first dolphins. One method of determining what forms animals had in the past is to look for their remains in rock formations, which can be dated. Thus the study of fossils can give us an insight into evolution.

Most times, when an animal or plant dies, it disintegrates and the products are recycled. If this were not so the earth would not be the place of vigorous growth and change it is today. However, there are rare circumstances when an animal dies, in a bog, say, and is preserved for sufficient time for salts to be deposited in the bones. These remains become covered with sand and sediment which eventually fuse into rock. When the rock is cracked open, perhaps millions of years later, we see the shape of the animal as a fossil.

In the sea the chance of this happening to a large structure is exceedingly remote. When it does occur the result will not be revealed to the eye of man unless the rock is subsequently raised above sea level and then is worn away, or changed in some way.

In the case of dolphins very few fossils have been found to date. So we must also look to other sources for tracing back their evolutionary development. One way this can be done is to examine the changes that take place in the developing embryo. For complex changes which take place from the moment the egg is fertilized until the baby dolphin is born are considered by many scientists to reflect the evolutionary path of the species.

Even when all this scientific evidence is taken into consideration there is still much left to speculation. So what kind of picture can we put together?

Firstly we must travel back in time, 30 to 40 million years, to a small four-legged mammal that lived beside the sea. There was food available on the land of course, but there was also a great abundance of food in the sea. Without any tools, animals probably waded into the sea and grabbed some of this food in their mouths. This method of feeding proved satisfactory and certain animals got more of their food from this source. Those developing features which enabled them to move on the land and in the water were the most successful, and flourished. Gradually, over countless generations, they completely adapted to a marine environment, yet still retained one feature which they had acquired on the land: they were warm blooded. It was this characteristic which had already set them at the top of the evolutionary tree on land. For by controlling their body temperature and keeping it more or less constant they were able to be active at all times, unlike the reptiles that needed to warm-up

This dolphin mosaic at Olous on the island of Crete was laid as a floor in an early Christian Basilica.

The skeleton of a Bottlenose dolphin in the Galapagos Islands reveals the numerous vertebrae in the backbone. The muscles attached to their backbones enable dolphins to develop extremely powerful thrusts with their tails.

in the sun before they could move fast.

The mechanics of moving efficiently through water were, of course, completely different to those of moving on land. The problems of swimming had already been resolved in evolutionary terms by the fishes. It is perhaps not surprising that the mammal mutants which became the most successful in the sea were those that adopted a basic fish shape. The major propulsive force in most fishes is derived from the tail by side-to-side movements of the backbone. The land mammals evolved to have a far greater strength and flexibility by a bending movement of the spine in the line of the body. Thus, although the fish shape was adopted, it was accompanied by a unique up-and-down movement of the tail which is characteristic of all dolphins today.

19

Having evolved a fish shape, an obvious question which arises is why not go all the way and change the lungs for gills, thus enabling the archeocetes, as the first dolphins are called, to stay underwater permanently? To answer that we must return to the point about dolphins being warm-blooded. There are two things to remember in this context: firstly, a very large area of contact is needed for the oxygen and carbon dioxide to diffuse in and out of the blood; secondly, water is an extremely good conductor of heat. So if water was passed over a gill-like organ in the dolphin it would drain away an enormous amount of heat which would have to be replaced; and to do this would require an impossibly high intake of food. The dolphins therefore, like all other mammals, stayed air breathing.

Thus through this speculative process of evolution a very successful new life form appeared on our planet, with a number of essential features that set it apart from all other animals:

It was roughly fish shaped.
It was totally aquatic (i.e. it spent its entire life in water).
It was air breathing.
It was warm-blooded.

Approximately four-fifths of the earth was covered with water, and in almost every way the first dolphins were superior to all other life forms in the sea. So the scene was now set for some rapid and dramatic changes in evolutionary terms.

The new arrivals spread across the oceans of the world and adapted to life in all seas. A new class of animals had arrived. They were the whales. They founded the order of animals we now call the cetaceans.

At the present time the cetaceans, or whales, can be divided into two quite distinct and separate sub-orders. These two exist because they evolved to specialize in the pursuit of different types of food. One group went after the tiny particles of food suspended in the water. To capture these they developed huge mouths which were capable of gulping large volumes of sea water and then straining the food particles from it. One of the richest sources of such food is the tiny shrimp-like creatures called krill which are found in huge quantities in the polar regions.

The mouths of whales that go after such food are lined with long horny plates which are covered in bristles. These strips, called baleen, were once put to a wide variety of uses: one of the most common being as stiffeners in garments such as corsets. Nowadays plastic is used in place of baleen.

The masses of baleen which lined the mouths of the whales were like giant whiskers, and the species which had this feature were given the name *Mysticeti* – which literally translated means 'moustached whales'.

Small marine animals without teeth would be vulnerable to attack. This was probably the reason for the evolution of the mystecetes as large animals that would have little to fear from sharks, simply because of their huge bulk. Indeed the mystecetes are by far the largest animals ever to have lived on earth, with the Blue whale at the top of the size league, reaching over thirty-two metres in length and weighing as much as eighteen London double-decker buses. It is interesting to note that the largest of the true fishes in existence today, the whale shark, which reaches a length of about eleven metres, is also quite harmless to man because it too eats only plankton.

The whales that went for bigger prey retained their teeth and form a sub-order of the cetaceans called the *Odontoceti*, which simply means 'toothed whales'. The largest of the toothed whales is the Sperm whale, which has teeth which are conical in shape and about ten centimetres long. The smallest of the odontocetes is the Common porpoise, which grows to a maximum length of about two metres and has small interlocking spade-like teeth.

Deciding what is a dolphin and what is not a dolphin can cause a great deal of confusion. Firstly, because the classifications of cetaceans is complicated – even the experts on this subject (called taxonomists) differ amongst themselves. Secondly, because the Americans tend to call all dolphins 'porpoises'. And thirdly, because there is a true fish which is also called a dolphin, or a dorado, and has the Latin name *Coryphaena hippurus*.

In this book we shall regard all *Odontoceti* (Toothed whales) as dolphins, with the exception of the *Physteridae* (the Sperm whale family) and the *Ziphiidae* (the Beaked whale family). This leaves us with the following families: *Phocoenidae* (porpoises), *Delphinidae* (dolphins), *Stenidae* (White dolphins), the *Plantanistidae* (River dolphins), *Monodontidae* (White whales) and *Globicephalidae* (Pilot whales).

Each of these families is divided by taxonomists into smaller groups called genera. Finally each dolphin is designated as a specific species. Thus Donald, the wild dolphin I studied, belonged to the species *Tursiops truncatus* (common

The dorado fish is also called a dolphin fish. It is a true fish, i.e. it is cold blooded and has gills, but its name can cause a great deal of confusion – especially when it appears as 'dolphin' on a restaurant menu.

name 'Atlantic Bottlenose dolphin'). His genera is *Tursiops* (Bottlenose dolphin) of the family *Delphinidae*.

If you find all of that too confusing then let us just agree that for our purposes dolphins are small whales, and leave it at that.

3

Dolphins Around the World

WE HAVE just seen that the members of the dolphin family can range from the 1.5 metre-long Common porpoise to the 9 metre-long Killer whale.

Let us now look at some of the dolphins that lie between these two extremes. We cannot examine all of them because we are not sure exactly how many there are. Some species have yet to be described in detail since nobody has seen a well preserved specimen. Indeed, some dolphins are known only by study of a single rotting corpse. The problem is, of course, that dolphins spend their lives at sea and so are difficult to observe when alive. When they die their bodies sink and are lost. Occasionally, corpses are washed ashore. But the bodies of those species that inhabit deep water are unlikely to end up on the land; and as no extensive bottom trawling for fish is done off the continental shelfs, the dead specimens are not recovered by fishermen.

The dolphins best known to the general public belong to the genera called *Tursiops*. *Tursiops* are the Bottlenose dolphins. They are found in all of the temperate and tropical oceans of the world – different species inhabiting certain areas. Those that swim around Britain are of the species *Tursiops truncatus*. They grow to a size of 3.5 metres and were first given the common name 'Bottlenose' because the head and upper jawline resembled the shape of early wine bottles.

Another dolphin which is well known is the Common dolphin of the genera *Delphinus*. The Common dolphin is a most beautiful animal to see in the wild state. The colour markings along its flank range from brown to yellow to cream, but on death these quickly fade to a nondescript grey.

The species found in the Mediterranean, *Delphinus delphis*, was well known to the ancients and it is the dolphin depicted in classical architecture and coins. One of the most colourful images was painted in 1500 BC. It formed part of the fresco in the queen's bathroom in the temple at Knossos on the island of Crete, and can still be seen today.

The Common dolphin is slender and grows to a length of 1.8 to 2.5 metres. Two elongated lens-shaped light areas on each flank are prominent, giving a figure of eight appearance which can be distinguished easily from a distance. The dark pigment along the back has a characteristic outline which has led to the name 'Saddleback dolphin' in the United States of America.

Sometimes huge schools of dolphins are encountered in the Mediterranean; the sea can appear to be alive with them. They are not confined to this area, however, and large schools are also encountered in other seas such as those off South Africa and New Zealand.

The Common porpoise, or Harbour porpoise as it is sometimes known, prefers shallow coastal waters at certain times of the year. They will enter river estuaries and are occasionally encountered well upstream. According to an old rhyme, when porpoises were seen in the River Thames in London, it was a sign of bad weather. Although less common than in former years, the Common porpoise can still be found, sometimes in large schools, or congregations, in the temperate and cool coastal waters of the northern hemisphere. The porpoise is characterized by its tubby rounded head and its triangular dorsal fin in the middle of the back. The back is black in colour, and this is usually all that can be seen unless they are in a playful mood. Then the white or cream belly may be revealed briefly as the porpoise leaps out of the water. Common porpoises have spade-shaped teeth.

A slightly larger porpoise that is commonly seen in the North Pacific is Dall's porpoise – *Phocoenoides*. It is easily recognizable by a large white egg-shaped area on the side beneath the dorsal fin, which may also be tipped white. Dall's porpoise often plays in the bow waves of boats. This may be derived from their habit of accompanying large whales. It is possible that they have known for a long time how to hitch an easier passage on their migration routes from California and Japan to the sub-Arctic coasts.

This little porpoise seldom leaps clear of the water like many members of the dolphin family. When travelling at speed close to the surface it sets up a coxcomb of spray which enables it to be identified often before its characteristic black and white markings can be seen clearly. Dall's porpoise is

24

A Bottlenose dolphin.

believed to be one of the fastest of cetaceans despite its relatively small size. A sea captain accurately timed the period it took for a Dall's porpoise to overtake his vessel. On the basis of his observations he calculated that the porpoise's speed was in excess of 55 kilometres/hour (35 miles/hour).

As its name implies the Black Finless porpoise *Neophocaena phocaenoides)* has no dorsal fin. It is a slim, lively species that grows to a length of only 1.5 metres. It is widely distributed and has been seen 1,600 kilometres (1,000 miles) up the Yangste River in China. The Black Finless porpoise doesn't seem to mind whether the water is hot or cold, for it has been spotted off the

The author at the Minoan temple at Knossos in Crete where a dolphin fresco was painted to decorate the Queen's bathroom in 1500 BC.

coast of Japan when the temperature was an icy 5°C, as well as in near tropical waters where the temperature can be as high as 25°C.

Some dolphins are occasionally found in rivers, but they spend most of their time at sea. There is, however, one family of dolphins which spend most of their lives in fresh water. They are the unusual River dolphins. The family name of these fresh water dolphins is *Plantinistidae*, and they are characterized by their long narrow beaks which are used for probing the muddy river beds. It is impossible to see very far underwater in most rivers because of the mud in suspension. So the fresh water dolphins have little use for sight. However they are experts at navigation and find their food with sonar. The Indus River dolphin (*Platanista minor*) is completely blind and has only a tiny vestigial eye, which has no lens. A similar species (*Platanista gangetica*) lives mainly in the muddy Ganges River and is often found in waters set aside for Hindu religious observances, so that bathers sometimes have the close company of dolphins during their worship.

Other fresh water dolphins live in the Amazon River in South America and in the large rivers of India and China. The Chinese River dolphin (*Lipotes vexillifer*) is also known as the White Fin dolphin because it can be spotted when its small light coloured dorsal fin breaks the surface.

The Indus River dolphin has a ridge in the centre of its back and lacks a prominent dorsal fin. Like most of the other river dolphins, however, it has relatively large paddle-shaped flippers which are used for feeling its way along the river bed, as well as acting as hydrofoils for steering. The river dolphins are also able to turn their heads from side to side to a much greater degree than most of the dolphins that live in the sea. This probably makes catching their prey easier because of the lack of space in their home waters, where they would rely less on speed and more on agility to catch fish.

The only other dolphins that have the ability to turn their head from side to side live in a totally different environment from the tropical waters of the River dolphins. I refer to two unique animals: the Narwhal (*Monodon monoceros*) and the Beluga or White whale (*Delphinapterus lencas*). Although bodies of both these species have been found washed up on the beaches of Britain, they are seldom seen outside the polar seas. They belong to the same family, and neither has dorsal fins.

The Narwhal is the only naturally occurring one-horned mammal in the world, and it probably gave rise to the idea of the mythical unicorn. It is a fearsome-looking member of the dolphin family and can weigh up to 1000 kgs (one ton). A fully grown male may reach 4.5 metres in length, exclusive of the tusk, which can add another three metres in the oldest animals.

The hollow tusk, which is twisted in a tight spiral, is a modified tooth that grows along the same axis as the body. It is very tough and will polish to a creamy gloss like elephant ivory. Nowadays Narwhal tusks have value as curios. Some people have fashioned them into walking sticks; sailors and eskimos carve them.

Nobody is sure what benefit the Narwhal gets from its magnificent tusk, which occurs only in the male. It has an obvious use as a defensive weapon but would be of no value for spearing prey, for the Narwhal has a relatively small mouth and feeds mainly on cod and squid. It has been suggested that the tusk might be used for making a hole in the ice through which the Narwhal can breathe. Although Narwhals have been observed with their heads sticking through the ice, nobody has reported seeing one using its tusk as an ice chisel. So for the present time the function of the Narwhal's tusk remains one of the sea's many mysteries.

The Beluga or White whale also inhabits the polar seas, but it has no tusk. The calf is grey/brown when born and turns blue/grey, usually in the second year. The all-over cream or white colour of the adult is attained gradually, over five or six years, by which time the animals usually reach a length of 3.5 to 4.25 metres. The colour acts as camouflage when they are resting in the ice floes; but in open water these beautiful whales can easily be seen and recognized, especially when they swim ahead of the bow of a ship. They are playful and agile. Belugas can turn their heads from side to side because the vertebrae in the upper backbone are not fused as they are in most of the other dolphins. Whalemen often called them 'sea canaries' because of the noises they made. They are gentle creatures that feed on prawns and small fish. They make long migrations and occasionally swim far up rivers. In 1966, a young Beluga swam up the River Rhine through Holland into Germany.

The Beluga spends much of its time in the Arctic where it must find an opening in the ice to breathe. Like all wild animals it must be alert in this harsh barren region, for polar bears have been known to wait by a breathing hole in the ice. When the Beluga surfaces, it is stunned by a mighty blow from the bear's paw. The bear then hauls its victim out on to the ice.

Another large dolphin that prefers cooler waters, although it refrains from venturing deep into the Arctic and Antarctic, is the Pilot whale (*Globicephala melaena*) which is known as the Caa'ing or Calling whale in the Orkneys. The first of its common names can be traced to the belief that it guided (or piloted) fishermen to shoals of herring. The second common name was derived from the fact that these dolphins often produce a great variety of sounds.

A different species of Pilot whale (*Globicephala macrorhychus*) which is slightly smaller is commonly called the Pacific Pilot whale. It prefers warmer waters and is often found in the company of Bottlenose dolphins and Pacific White-sided dolphins. Like the Pilot whales from colder climes, the Pacific Pilot whales make a lot of sounds that can sometimes be heard clearly above water.

The most magnificent of the dolphins is the Orca (*Orcinus orca*), which also bears the name of Killer whale. It is the largest of all the typically dolphin-shaped whales, and the males can reach nine metres in length. In the wild a mature male is characterized by its tall dorsal fin, which stands 1.5 or 1.8 metres high and is as upright as a flagstaff. The dorsal fin of the female is recurved and smaller.

The name Killer whale is a controversial one because it implies menace, so those who have studied these beautiful animals prefer to call them Orcas. Orcas hunt in packs and have been known to feed on seals and small dolphins; an attack on a Blue whale has also been recorded.[*] Observers have noted that a school, or pod, of Orcas will act like a team when hunting. When going after seals which are resting on an isolated rock, for example, some of the Orcas will go to one side and thrash the water with their tails as they swim close inshore. This frightens the seals who panic and rush into the water on the opposite side of the rock, where the remainder of the Orcas lie quietly in wait, just below the surface.

It has also been reported that Orcas have rocked an ice floe with men on it, in an attempt to dislodge them into the water. However, if this was really the case, the men survived to tell the tale. Good yarns like this often served to enhance the reputation of the first explorers of the Arctic region, who were undoubtedly very brave men. But the story has now been brought into question on three counts by those who have studied Orcas in captivity and in the wild.

[*] *National Geographic 155* pp 542–545 (April 1979)

The Beluga which inhabits arctic seas has additional vertebrae in the neck which enable the head to move from side to side more than in most other dolphins.

Firstly, Orcas eat mainly fish. If there is plenty of this food around, it seems contrary to the nature of the dolphin family as a whole to attack other mammals not of their own species. Secondly, Orcas are sometimes accompanied by small dolphins and porpoises when swimming as a school in the wild; the different species appear to live and swim in harmony with one another. The third and most significant point is that no member of the dolphin family has ever made an unprovoked attack on man – as far as we know. On a number of occasions dolphins have indeed helped to save the lives of people in distress at sea.

Thus there is a conflict of opinion on the nature of the *Orcinus orca*. Should the name Killer whale be completely abandoned? Those who have seen these magnificent giants in captivity will wonder at the power they have. At the

30

The erect dorsal fin of the male Killer whale enables it to be identified easily. The Orca in this picture is larger and leaner than the boat in which the scientist studying him is standing.

same time we must marvel at the gentleness they show towards the humans around them. This is especially remarkable when one considers what effect imprisonment in a small pool must have on a large animal that would normally travel vast distances in the open sea.

A savage bite from an Orca could kill a man almost instantly. The fact that divers have often swum safely with them identifies a very special quality that is characteristic of the entire dolphin family. For all dolphins seem to have a special affinity with man, and many people have a great love of dolphins. It is as if there is a link between them. Nobody can say exactly what it is, but it is there nonetheless. Part of the magic of the dolphin is the mystery of that special bond.

4

The Life of a Dolphin

DOLPHINS are superbly adapted to a life in water and this becomes apparent from the moment they are born. Land mammals are normally born head first, but dolphins leave the womb tail first. This is called a breach birth.

Inside the womb, the baby grows in a fluid-filled cavity and is connected to the mother via the umbilical cord.

The later stages of pregnancy are usually obvious in humans because of the increase in size of the abdomen, and the mother-to-be becomes less agile. This is not so with dolphins. As the foetus inside develops, the mother retains her smooth lines and shows no signs of impaired activity until the moment the baby dolphin is born.

When any mammal gives birth many changes must take place in a short space of time in both the mother and her offspring. Even if everything goes according to plan, it is usually a very exhausting period for the mother. And there are times when difficulties arise. It is for this reason that most Western mothers-to-be go into hospital to have their first child; mothers who have their babies at home call upon the assistance of a midwife. An ability to call upon help at times like this is one of the benefits of a civilized society, but it appears that dolphins enjoy the same kind of advanced social behaviour. Another female dolphin is usually present to help the mother through the difficult time of birth. The dolphin 'midwife' will help the calf to the surface to take its first breath. She will also keep any curious males at a respectable distance, and they in turn will deter any sharks or other predators attracted by the blood which is always present when a mammal gives birth.

The newborn dolphin usually weighs about fifteen per cent of its mother's weight and can be almost half her length. The calf is well developed and able to swim from the start of its life. It stays close by its mother and in a short time is able to swim at speed.

For the first few days after birth the mother and calf appear to be connected by an invisible elastic thread, for they swim in close harmony together and rise simultaneously to take a breath of air. However, the youngster soon becomes more adventurous, but never strays too far from the protection of his mother.

The calf grows very rapidly on a diet of milk which is thick and nutritious. It contains more fat and less water than the milk of land-based mammals. When the calf is feeding a slit on the underside of the mother opens and the nipple is protruded. The calf's tongue forms a tube when it is pressed against the roof of the mouth during suckling. This allows the milk to flow into the stomach without the entry of sea water.

The calf cannot suckle for very long without the need to surface to breathe. So feeding has to be efficient and quick. Nature has solved this problem by providing the mother with a duct in which the milk collects. This milk is squirted into the throat of the calf when it suckles. When the calf stops suckling the nipple is withdrawn into the slit, and the mother's body once again resumes its smooth line – which offers the minimum of resistance when moving through the water.

For the first six months of its life the young dolphin lives exclusively on a diet of milk. After that it learns to catch fish and starts to feed itself as the supply of milk from the mother slowly dwindles. Even so, the mother may continue to suckle her calf intermittently for up to eighteen months.

Little is yet known about the social life of dolphins in the wild. A young dolphin will sometimes be seen swimming in the company of two adults, but how the three are related is not clear. It is often difficult to tell the sex of a dolphin without close inspection, so it is just possible that one of the trio is a father keeping a protective eye on his offspring. The group of three may form part of a school which can vary in size on both a daily and a seasonal basis. Sometimes a school of dolphins will stay in a given area of the sea for prolonged periods. At other times, it may join a huge congregation of dolphins and migrate to new waters.

Two young Killer whales play together in the wild. The one in the foreground is only a few weeks old and is being watched over by his three year old sister. The sea is a hostile environment for a mammal to be born into and unlike a human baby the dolphin must be able to move swiftly from birth.

The young dolphin spends its entire life on the move. Just how far it travels and how long it spends in different areas of the sea will depend upon what species it is, and on the food supply. If it is a Bottlenose dolphin, it will tend to spend part of its time in coastal waters. If it is a Spinner dolphin, it will pass more of his days in deeper water. Some dolphins, such as the Beluga, limit their travels to those areas of the world where the water is cool. Other dolphins – like the largest of them all, the Orca – probably roam from the equator to the pole.

The young are born in the warmest waters frequented by their own particular species. The mother and her calf form part of a school or pod of dolphins that will vary in number and may contain more than one species.

34

The youngster learns about the way in which members of the group work together.

There is close contact between the mother and her offspring. Like most young mammals, dolphin calves are playful and there are times when their games get out of hand. When this happens a mother will use her tail to slap the offender. Like all youngsters, baby dolphins have to be taught who is master and to follow some rules of behaviour. However, dolphins are very affectionate and caring, and the mother and her rapidly growing infant swim and play together like two close companions.

We know that the largest whales – the mystecetes – feed intensively in cold, plankton-rich waters for part of the year. This food is stored as fat to fuel their long migrations to the warmer waters of their breeding grounds. During this part of their year, which may last for six months, they take in very little, if any, food, and must live on their reserves.

Food is always available for fish-eating dolphins, however, and they probably feed every day. I say probably because nobody has yet been able to study the eating habits of a wild dolphin in the sea throughout the course of a year. An active, growing dolphin may eat up to 15% of its weight in fish each day. This could reduce to about 5% in an adult. Thus a 180 kilogrammes (400 pound) adult dolphin will eat something like 9 kilogrammes (20 pounds) of fish per day.

It is reasonable to assume that like most wild animals the dolphins will have seasonal variations in both the quantity and type of food they eat. The type of food will also vary according to the habitat of the various species. So the river dolphins will eat fresh water fish and some may probe the muddy river bed for bottom-dwellers, whilst the fast-swimming Common dolphin will feed on shoals of pelagic fishes that swim close to the surface in the open sea as well as diving deep to feed on lantern fish and squid.

Fresh fish contains vitamins which are essential to the health of a dolphin. Because some nutritional value is lost when fish are killed and gutted or frozen, vitamin supplements are incorporated in the diets of dolphins kept in captivity. They are usually fed on fish that has been kept in cold storage. It is thawed before it is given to the dolphins, but as it is not cooked it is essential that only high quality fish is used.

Dolphins have three compartments to their stomachs, which resemble those of cattle. In the stomach the flesh from the fish is partially digested and stripped from the bones, which are regurgitated, leaving a soft mass that

35

passes into the narrow intestine. Foreign objects thrown into a dolphin pool can have fatal effects. Rubber balls, torch batteries, keys, and all manner of unlikely things have been discovered in the stomachs of dolphins during post mortem examinations.

In the wild, dolphins sometimes act as a team when feeding. They will form a line and herd a shoal of fish towards a cliff face. When the fish are all bunched together and have no way of escape, the dolphins move in for an easy meal.

The young dolphin learns how to catch fish and to turn them in its mouth so that they can be swallowed head first; for dolphins do not chew their catch. They use their teeth for capture only. The teeth are all roughly the same shape and interlock like a very loose zip fastener.

Dolphins do not feed entirely on fish. The weaned dolphin finds out how to forage on the sea bed, and soon its diet is supplemented with squid, shrimps and even small sharks. The depth to which a dolphin will dive for food depends upon the species. Obviously those who inhabit the open seas will have to go deeper than those in shallow water.

A large brain is the feature which characterizes the so-called higher animals. When trying to get an insight into the daily life of a dolphin, a comparison of the size of different parts of the brain which are known to be associated with specific bodily functions can help us to a greater understanding. For instance, smell plays an important role in mating, hunting and survival in some animals. In such species the part of the brain associated with this function is relatively large. In dolphins, though, it is almost non-existent. The conclusion drawn, therefore, is that dolphins have a very poor sense of smell. This is not very surprising when one considers how little time the dolphin actually spends in the air which carries scents and smells.

The region of the brain that involves the decoding and analysis of sound is very large in the dolphin. It has often been stated that dolphins "see with sound". But when we talk about sound vision it becomes difficult to understand just what the dolphin "sees". When dolphins are hunting they emit a range of very high frequency notes and listen for the reflected sound, or echo. From this they are able to tell the size of a fish and how far away it is. A major part of the echo will come from the swim bladder – an air-filled sack inside the fish which enables it to regulate its buoyancy. Because air has a different density to the rest of the body it gives off a different sound signal. The dolphin

Occasionally a dolphin is born with very little pigment in its skin. The young, cream-coloured, albino Common dolphin in the top centre of this photograph was often spotted off Gibraltar.

will therefore "see" inside the fish as well as outside. Thus the information processed in the dolphin's brain will be more like an X-ray than a photograph taken with light.

The dolphins' ability to swim at great speeds give them an obvious advantage when hunting. An accurate speed reading has been obtained for a young female dolphin, just 1.6m long, who was captured in the Black Sea and then released back into the fast moving school after a line had been attached to her tail. From the speed with which the line ran out, it was calculated that her speed was 44.4 kilometres/hour (27.55 miles/hour). Common dolphins have been clocked doing between 55.2 – 58.8 kilometres/hour (34.5 – 36.8 miles/hour) when riding the bow waves of destroyers. This may not be a true free swimming speed, however, because the dolphins are pushed along by the waves in front of the ships. Nonetheless Common dolphins are amongst the swiftest of all the cetacea.

37

Dolphins swim with up-and-down movements of their tails, not side-to-side motions like fish. Their superb swimming ability still remains somewhat of a mystery. The skin plays an important role because this is the part that is in contact with the water and produces drag. Dolphins are not covered with a layer of dead skin cells like the human epidermis. They continuously secrete an oily substance which slides off the skin and is carried away in the water, reducing friction in the boundary layers between the water and their skin. As a result a dolphin's skin has a unique feel when touched. It is not scaly or slimy like a fish. It is not dry like that of a human. It is firm yet yields slightly to touch, is very smooth, and feels neither warm nor cold.

Submarine designers have looked at the dolphin with envy. For despite all of man's scientific knowledge, he cannot match the dolphin's incredible speed with any machine developing the same power as the tail. Indeed dolphins seem to defy the laws of hydrodynamics. And if you have ever tried treading water, even with fins on, you will appreciate just how much thrust dolphins develop when they jump high into the air.

Dolphins usually reach sexual maturity at about seven years of age, but there are few external signs of the changes taking place because the organs associated with reproduction are hidden. The sea is a very hostile environment. Any part of the dolphin that protrudes must be tough, or it is likely to be damaged in the hurly-burly of a vigorous life. Also there is a need to conserve heat. Thus most parts of the animal are covered in a layer of blubber. And finally, any part that projects into the water will cause drag and reduce the dolphin's speed.

For these reasons the parts of the body which normally enable the sex of an animal to be determined are kept well out of sight. When fully grown the female is smaller than the male, and may be a little less brightly coloured.

The one species of dolphin where the male and female can be easily spotted is that of the Orca. When he is fully grown the male has a magnificent dorsal fin that is almost symmetrical and rises upright 1.5 metres from his back. The female's dorsal fin is much smaller and is recurved.

Dolphins engage in courtship games and rub against one another. The penis is composed of a fibro–elastic tissue and can be flicked out rapidly with the erector penis muscle. During mating, the erect sensitive penis is inserted into the vagina of the female underwater when the two come together abdomen to abdomen. General observations would indicate that mating

It is difficult to tell the sex of most dolphins because the organs associated with reproduction are normally tucked away to maintain a streamlined shape. The presence of a slit in the abdomen about a third of the way from the tail to the head enables this Bottlenose dolphin to be identified as a male.

occurs randomly within a school of wild dolphins.

Bottlenose dolphins distributed along the Atlantic coast of America and Europe give birth between February and May. Mating also occurs during these months and this implies a gestation period of about one year. It is possible, however, that the period of pregnancy can vary because of a process known as delayed implantation. This is nature's way of ensuring that the young are born at the appropriate time of the year, regardless of when mating occurs. The fertilized egg is stored inside the female and does not become attached to the womb and grow until the appropriate time – corresponding to a spring birth – is reached.

During its period inside the mother the growing foetus receives its supplies of food and oxygen via the umbilical cord. This cord breaks at the moment of birth and the newborn calf makes its way instinctively to the surface, where contact with air stimulates it to breathe. Dolphins give birth to a single offspring.

In a balanced world, for every animal that is born another must die. If they survive the hazards of early life and the other dangers in the sea, some may live for about forty years. A dolphin has only one set of teeth in its life and after about twenty years they show signs of ageing. In a very old dolphin the teeth may be worn down to stumps.

At the end of its days the dolphin's blow hole relaxes. The lungs fill with water, and the dolphin sinks slowly into the depths.

5

Dolphin Intelligence

EXACTLY how and when man first made his appearance on earth is still the subject of much discussion. The so-called "missing link" which will connect primitive man to the rest of the tree of evolution has yet to be found; so too have the missing link or links which connect present-day dolphins to the earliest air-breathing, fish-like creatures called archeocetes.

Scientists are still hunting for fossils which will help them put together the complex jigsaw puzzle of the evolution of men and dolphins. The evidence available so far indicates – in very round figures – that human-like beings have been on the earth for about four million years. It is now generally accepted that animal life first came into existence in the sea, and that some creatures migrated to the land. Eventually animals, which we now call mammals, evolved and were very successful relative to the reptiles that earlier dominated the land.

A study of the development of a baby dolphin, from the fertilization of the egg to birth, coupled with the analysis of the bone structure of the adult together with other evidence, leads us to believe that dolphins are derived from land mammals. Thus one of the proposed connections between man and dolphins is that we had common ancestors many millions of years ago.

Since those uncertain prehistoric times, changes have taken place. One of them, about which we can be absolutely positive, is that present day dolphins are the animals with the largest brains, on a weight for weight basis, in the sea. And man has the largest brain, weight for weight, on the land. Because of this similarity and their common ancestry dolphins have been described as man's cousins in the seas.

The Bottlenose dolphin has a brain which is approximately the same size as that of a man. Does this picture show an encounter between two intelligent beings? At present expert opinions differ on this subject.

Although nobody can deny the similarities in the weights of human and dolphin brains, there is much discussion and disagreement about one of the characteristics which we think is associated with brain size. And that is intelligence. Although admitting that dolphins are delightful and entertaining creatures, some scientists maintain that all the evidence we have so far indicates that dolphins should be rated along with dogs and chimpanzees on the intelligence scale. Others, however, postulate that dolphins have a potential for intelligence at least as high, if not higher, than that of man.

Strange as it may seem the answer to this dilemma could be that both groups are right. It all depends upon how the word intelligence is defined. If intelligence is related to an ability to devise, manufacture and manipulate

Dolphins are often very curious – especially towards new sounds – as the author found out when he rattled a folding anchor whilst diving with a friendly wild dolphin off the coast of Brittany.

tools, then man is undoubtedly superior to dolphins. If though, the dolphin has deployed its mental capabilities in a completely different direction, then a man, who thinks in mechanical terms, may not be able to understand the working of the dolphin's mind.

The basis of modern science is that evidence must be produced in the form of a repeatable experiment before a proposition becomes acceptable. So all we can do at this stage is look at the evidence and draw our conclusions.

One of the reasons why intelligence is ascribed to dolphins is the speed with which they learn tricks when in captivity. Virtually all of the actions seen in a pool are also performed spontaneously by dolphins in the wild. However, they do respond very quickly and with great precision to instructions. These are sometimes conveyed by whistle – a dog whistle is often used and the audience do not hear the sound – but more often the dolphin trainer will use arm signals. Frank Robson, a dolphin expert in New Zealand, claims to have trained dolphins simply by concentrating his mind on the tricks he wished them to perform; in other words, by telepathy. Interestingly he did not reward his dolphins with food; they did their tricks to please him.

A very close bond develops between the trainers and the dolphins in their care. All the trainers I have met have deeply loved their dolphins. If a dolphin misbehaves itself in a show the trainer does not have to beat it, or stop feeding it. The trainer has merely to show displeasure by turning his or her back on the animal and walking away from the pool. The dolphin is so upset by the apparent withdrawal of affection that it will usually do what is expected of it the next time the signal is given.

The trainers do not have it all their own way all of the time, however. There are a few occasions when the dolphins go on strike, usually led by the dominant dolphin in the pool. If one decides that his team has done enough, it stops performing and prevents the other dolphins from doing so.

It is this kind of situation, and the repetitious nature of dolphin shows, that has caused many trainers great concern. They get the feeling that the dolphins are highly intelligent and that keeping such animals in captivity is morally wrong. So much so that they feel compelled to quit their jobs despite the love they have for their performing partners. Two technicians who worked with dolphins used for scientific research in Hawaii felt this urge so strongly that they returned two captive dolphins to the sea. The men did so knowing that they were breaking the law and would have to face the consequences.

A number of articles have appeared indicating that dolphins have been trained for military purposes. I have seen a film of a Pilot whale in a test situation putting a grappling device on a warhead resting on the seabed. And I have no doubt that attempts have been made to train dolphins in anti-personnel tactics, i.e. to attack enemy divers. Stories have reached me of dolphins being armed with large hypodermic needles attached to their heads. The idea behind this eerie device is that the dolphin rams the swimmer! Compressed air, or carbon dioxide, is released through the needle killing the unfortunate victim almost instantly. Despite these stories, however, I do not have any reliable evidence that such devices have been actually tried and tested – although I have little doubt that such an idea has been conjured up in

44

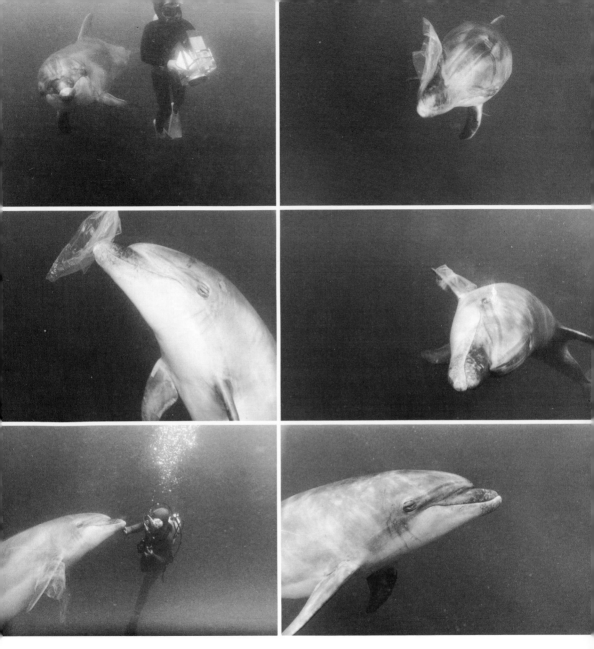

Dolphins initiate their own games. A wild dolphin called Jean-Louis turned up in front of the camera with a plastic bag to play with when the author was making a television film off the French coast.

The author and a wild dolphin at close quarters. Friendly wild dolphins appear to have a mischievous sense of humour which makes conventional research methods difficult.

somebody's mind. What he or she may have neglected to take into account, though, is the fact that dolphins are not normally aggressive animals and would therefore be difficult to train for tasks which do not come naturally to them.

Furthermore, all scientists who have worked directly with dolphins know that sometimes a dolphin will quite deliberately turn an experiment round, thus preventing the experimenters from getting the reproducible results they would from a dog, rat, or even a human. I think the reason for this may well be that dolphins appear to have a well-developed sense of humour. This ranges from getting a mouthful of water and spraying it over an unsuspecting person walking alongside a pool, to the deliberate mischievous antics of

Donald, the wild dolphin.

I have had an unconfirmed report that dolphins trained to stick limpet mines on enemy ships would do so nine times out of ten. But on the tenth time they would stick the mine on their own vessel. As a result, the idea of using dolphins to plant mines was abandoned. Whether it was the dolphin's sense of humour, or his intelligence, which led to this end I will leave you to decide. Whatever your conclusion, my own view is that man – with his ever present tendency to destroy both himself and all other life forms on our planet – should not attempt to corrupt the dolphin and convert a fellow creature into a killer. Perhaps the peace-loving dolphins have shown that they are too intelligent to be manipulated into living weapons of war.

If you think up a brilliant idea you will get no credit for it unless you can communicate the information to your fellow men. Communication, therefore, plays a key role in our understanding of what goes on in other people's minds. Dolphins use sound as their primary means of communicating with one another. Because of the very wide range of sounds they make, they have a potential for passing a great deal of information in a very short space of time. Indeed it has been suggested that if an alien intelligence were to attempt to communicate with life on earth, it might choose dolphins in preference to man.

Numerous attempts have been made to understand the language of dolphins; so far without success. Until we can find a way of communicating with them, whether via sound or other means, the complexities of their minds will remain a mystery.

6

Saved by the Dolphins

THERE are many stories about dolphins rescuing people. Not unnaturally these are often about people who find themselves thrown into the sea. The stories are frequently passed from person to person and it is difficult to get a first-hand account. However, when I was filming in Sierra Leone, on the west coast of Africa close to the equator, I came across a man who told me that his father had been saved from death at sea by dolphins.

The man, whose name was Moses Benga, took me to the place where his story started. It was a tiny village, mostly of mud hats, called Mama Beach. We sat together on an upturned dugout canoe in the shade under the coconut palms. A hundred metres away a group of about twenty fishermen and women were hauling a net from the sea, singing in rhythm as they pulled together. The sun burned down on the golden sand. We were surrounded by a crowd of wide-eyed children who listened attentively. For Moses Benga was well educated and very respected in his village, because he had an important post with the Sierra Leone Ministry of Tourism in Freetown.

Moses had previously explained to me how his grandfather, who was a hunter, had settled close to the spot where we sat and founded the village of Mama Beach. Moses' father grew up in the village. As wild game became scarce in the dense tropical forest around the settlement, he became a fisherman like most of the other villagers.

One day when Moses was fifteen years old his father, whose name was Zachious Benga, set out to sea with a group of other natives to fish for tarpon. His canoe, which was of soft green wood, had recently been hewn out of a single trunk. It carried only one person. Because of its unique shape it was called a Shelro canoe, after the Shelro tribe to which the Benga family belonged. It was a very hot day and the group paddled far out to sea. When they were at the fishing ground miles offshore, the fishermen separated. Each one was intent on getting the best catch. They scattered across the wide ocean.

The sun burned down and Moses' father was unaware that the heat was drying out the wood of his new canoe. As the afternoon came to a close the dispersed canoes started back for the shore, one by one. Moses' father was one of the last to leave the fishing grounds. He was paddling back to shore when suddenly there was a crack; a few moments later the canoe split in two from stem to stern. The fisherman found himself in the sea with the two halves of the canoe floating beside him. He managed to haul himself on to the larger of the pieces of wood. Balancing precariously, he shouted and tried to attract the attention of the black dots on the sea which indicated his companions heading homewards. But the old man's cries were lost in the wind and nobody turned round to see his plight. Tired and wet, he tried to paddle to the shore, but the current carried him down the coast and towards the open sea. As the sun went down, he could just see the land as a thin line on the horizon.

On the shore the villagers became worried about the fisherman who had not returned when the short tropical dusk turned to darkness. At first light the following morning there was still no sign of the old man or his canoe. Was he dead? The chances were that he was. Such incidents had happened many times before. Fishing in their dugout canoes was a dangerous business; the sharks made short work of anyone who was unfortunate enough to founder at sea.

But Moses' father was not dead. He lay prone on the upturned keel of his broken boat. It was a dark night and all around him he could hear splashing. He was terrified as he thought of the sharks waiting for him to lose consciousness and fall into the water. But he was determined to thwart their plans, and willed himself to stay awake and cling to his wooden raft. The night seemed to have no end, and all the time the old fisherman could hear the splashing nearby. When the pale light of dawn eventually spread across the sea Moses' father was surprised that he was still alive, and even more surprised to see that the splashing sharks around him were not sharks at all. They were dolphins. When the man realized who his companions had been he heaved a great sigh of relief. Hope flooded through him. With friendly dolphins for company there would be no shark attack.

49

As the sun rose high the dolphins circled the distressed man, giving him a renewed will to survive. Then, suddenly, the dolphins disappeared and the fisherman felt totally alone. He scanned the sea all around him. There was no sign of any canoes. The midday sun burned his back; he was getting weak and needed water. As the sun slid down towards the horizon he was still drifting hopelessly, and knew his chances of survival were now very small. He was conscious enough to appreciate the drop in temperature as the sun set and once again the darkness closed around him.

Moses' father knew that if he went to sleep he would fall from his float and would never get back on it again. His eyes closed, and just as he started to drift into unconsciousness he was startled into life again by the sound of splashing. The dolphins were back. For a time they filled him with hope. He talked to them as they slowly circled him. He could hear their breathing and felt as if he was with friends. But how could they help him?

As the night wore on the old man's weariness became overpowering, and he started to slide into sleep once more. But just as he was about to succumb to the desire, there was violent splashing around him. He was showered with water and woke up sharply. He realized that the dolphins were warning him not to go to sleep. And throughout the night they continued their vigilance, splashing him with water whenever he started to doze off.

At last the dawn came again, and Moses' father was still alive.

Miles away at Mama Beach village the fishermen were getting ready for another day out at sea, and the Benga family were preparing for the first stage of mourning which is practised three days after death.

Out to sea the fisherman, barely conscious, knew that he could not last another day without water. But he was comforted by the family of dolphins that kept him company. As the sun rose high in the sky and he prepared for death, the dolphins disappeared. But their presence had given him strength to cling on to his float until the end. As he was drifting into final unconsciousness, Moses' father was aware of a new sound. It was not the splashing of the dolphins, but the babble of human voices. He was too weak to move. Then, as if in a dream, he found himself leaving the sea and floating upwards. Black faces appeared out of the haze. People were talking to him, but the old man was unable to reply as his tongue was swollen inside his throat.

The men aboard the trawler that had discovered the fisherman on the

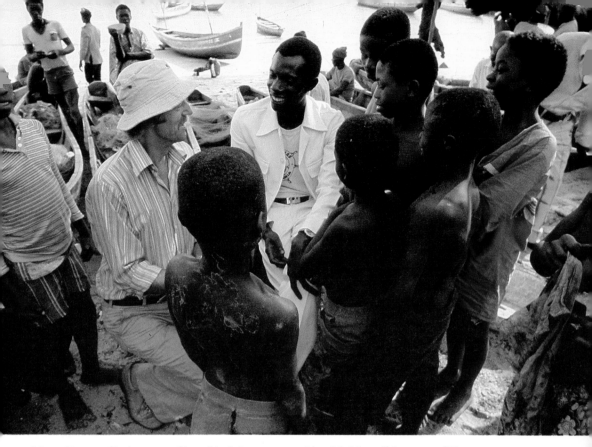

Moses Benga tells the story of how his father was saved by dolphins off the coast of Sierra Leone to the author and a group of entranced children.

remains of his upturned canoe lifted him gently aboard their boat, and were surprised to find that there was still life in the limp body. With great care they poured water between his parched lips, and nursed him back to consciousness.

The trawler carried Zachious Benga back to their port of Abidjan, where the old man rested and started to recover from his ordeal. There was no way in which he could communicate with his family in distant Mama Beach. So in his village the special ceremony for the seventh day after death went ahead.

When he was fit to travel the fisherman was moved to a hospital in Freetown; but still no word could be sent through to Mama Beach. In the meantime Moses' mother had sent messages to all of the relatives who could be contacted. She told them of the death of her husband. They were all expected to call at the house of the deceased man for the final ritual associated with his death. In black Christian communities in Sierra Leone this takes place forty days after death.

You can imagine the surprise when thirty days after he had been given up for dead, Zachious Benga returned to the village. His wife and his children wept with joy. When they heard of the way in which the dolphins had given the man the will to live, and had stopped him from going unconscious, they understood. For during the two generations the village had been in existence the fishermen had regularly seen dolphins on their trips to and from the fishing grounds. The men lived very close to the earth and sea, and had an innate understanding of the mysterious forces of nature.

A few days later I thought I might have to put the story Moses told me to the test. I was travelling in a very large native canoe in a rough sea, several miles off Mama Beach. Suddenly a huge wave swamped the vessel, which heeled over at an alarming angle. When I was convinced it was going to turn turtle, the canoe faltered and came upright before the next wave lifted it up on its crest. It was the closest I have been to capsizing in the many journeys I have made at sea. As we frantically baled out the water, I remembered Moses' story and hoped the dolphins were not too far away.

The story of a similar rescue by dolphins was widely covered by the media in November 1988. It concerned two sailors who were shipwrecked in rough seas off the coast of Indonesia. After their ship sank they were nudged and guided by dolphins throughout the night. The next morning the men reached the safety of a small island off the wildlife reserve of Ujung Kulong. Once ashore they were able to raise the alarm. As a result, rescue teams plucked another nine of the crew of the tanker *Elphina III* from the sea. Not all of those on board the vessel were saved, but the lucky ones who survived owed their lives to the dolphins.

7

Some Famous Dolphins

AS FAR as I know the story I have just told about the rescue of Moses Benga's father in Sierra Leone never made the headlines. The same was not true when in 1955 a Bottlenose dolphin turned up off the North Island of New Zealand.

Opo was a young female dolphin whose mother, it was believed, had been shot by a local youth. For when Opo appeared close to the small township of Opononi (hence the name Opo) she behaved as if she was seeking a lost parent, and would swim alongside small boats and nudge them.

Soon she allowed the occupants of the small boats to touch her. Gradually, the dolphin became bolder, and by the end of the year she would swim with the bathers in shallow water. One of Opo's favourite friends was a 13-year-old girl named Jill Baker. Jill was a good swimmer and was sometimes given a tow by Opo when they played together. Occasionally Jill would place small children on Opo's back and they would be taken for a short dolphin ride. The story of the friendly dolphin with a special love for children spread around the world, and thousands of people turned out to see her.

When Christmas came in 1955, which was in the height of summer in New Zealand, holidaymakers flocked into the sea to watch and touch the famous dolphin. They were not disappointed. Opo learned to play with a ball and flicked it high in the air with her tail. When the crowd cheered she would jump right out of the water. But Opo never let her high spirits get the better of her when close to young children. She always moved slowly in their presence so as not to endanger them. However, if a person was too rough and tried to grab her, the dolphin would signal her displeasure by smacking her tail loudly on the surface of the sea. Opo appeared almost every day throughout the idyllic summer.

The bones in the flippers of a dolphin correspond to those in the hands of humans. Is this dolphin extending the hands of friendship to the human interloper in his world?

But some people were concerned for the safety of the playful dolphin who had made their town famous. On 8th March 1956, a law was passed by the local council protecting Opo. On the following day the dolphin was found dead, cut to shreds and jammed in a crevice on the rim of a rock pool. She had been killed by a gelignite blast.

It is hard to understand why anyone would want to blow a dolphin to pieces, but sadly such people do exist. A wild animal can play safely with ten thousand people, but such is the power of man that it only takes a second for one person to bring the friendship to an end.

Opo was given a public funeral, and a statue was erected by the people who for nine months had enjoyed the magic of her friendship and the happiness she

This statue at Opononi, New Zealand, depicts Opo the dolphin who enchanted visitors to the seaside resort in the summer of 1955.

had brought to their lives.

When I visited New Zealand in 1981 I obtained a copy of a rare film of Opo. When I watched it for the first time I was surprised and delighted to discover how similar Opo's behaviour was to that of Donald. Although I did not know it when I first met him in 1974 – Donald was to change my life. I later made a television film about him. Thus Donald's mischievous antics were seen by millions and he too became a famous dolphin.

New Zealand has a special place in dolphin folklore. For another dolphin made his home there at the turn of the century. He was a Risso's dolphin, and he became famous because of his habit of escorting ships through Pelorus Sound – a stretch of water which required very careful navigation on the part of captains bringing their vessels from the North Island to Nelson in the South Island. Pelorus Jack first started his role as unpaid pilot in 1888, and continued until 1911. Many people came to watch him as he led the mail steamers safely into harbour. It is thought that Pelorus Jack was harpooned by a Norwegian whaling boat which was anchored off Pelorus Sound in April 1912, despite the fact that a special law had been passed to protect him.

As has been said, it is very difficult to tell the sex of a dolphin. It has always been assumed that Pelorus Jack was male; but there is a fifty per cent chance that this was wrong. If it had ever been proven that the friendly and helpful Risso's dolphin was a female, she would probably have been known as Pelorus Jill.

Accounts of friendships between people and wild animals provide a basis for good stories. One of the great storytellers of his age, over two thousand years ago, was the Roman, Pliny the Elder. One of the stories with which he entertained his audiences concerned the friendship of a peasant boy and a dolphin in the Mediterranean Sea. The friendship started when the boy, on his way to school one day, saw a solitary dolphin. "Simo, Simo," the boy called out over the water, and the dolphin responded and came to him. The boy offered the dolphin some of his bread, which was gratefully accepted. As the boy was very poor, he probably went hungry in order to feed the dolphin. However, he was rewarded for his generosity. Before long the two became friends and the dolphin would carry the boy across the bay to school. The friendship of the lonely boy and Simo lasted for years. But then the boy was taken ill and died. Each day after that the dolphin came to the beach to where he expected to pick up his passenger for the ride to school. When the boy did not turn up, the dolphin became sadder and sadder. Eventually the dolphin was found dead on the shore and the local people said he died of grief at the loss of his human friend.

Man has always been curious about the origins of life on earth. Nowadays the principles of evolution, as first outlined by Charles Darwin, are widely accepted. However, when our total knowledge was much smaller than it is today other explanations were found to account for the presence of dolphins on earth. The myths that were devised to explain it have now become part of our culture. But the myths had to incorporate what facts were known and they therefore contained elements of truth. Thus it is interesting, in view of the close association which many people now see between man and dolphins,

The story of the God Dionysus, who changed men into dolphins, is still told in the picture form, as on this modern plate made in Greece.

to look at the way in which the ancients explained the arrival of dolphins on earth.

One such mythological story is depicted on the famous Dionysus cup, which can be seen in a museum in Munich. The cup dates back to the year 540 B.C., and shows a crowned figure reclining in a dolphin-shaped vessel. Seven dolphins frolic around the vessel which has a vine, laden with grapes, sprouting from the mast. The ancient Greek potter who made the cup used

This ship and dolphin fresco from the island of Santorini, painted in 1500 BC, has been reconstructed in the National Museum of Athens and copies of it now decorate many homes.

his artistic skill to produce a beautiful object that was both useful and informative. For it describes the story of Dionysus, the Greek god of wine, who was later named Bacchus by the Romans. The legend tells how Dionysus, disguised as a man, was on a journey, when the sailors onboard the vessel on which he was travelling decided to sell him into slavery. Because he was a god Dionysus immediately became aware of the plot and decided to teach the treacherous crew a lesson. He caused the sound of flutes to fill the air and vines to sprout from the vessel. When the oars the sailors were using turned into snakes the terrified crew dived overboard.

Realizing that they were in the presence of a god the crew repented. So too did Dionysus, who changed them into dolphins to prevent them from drowning.

With such an explanation those in ancient times were able to understand why dolphins played around their boats; and why, if treated with kindness, they would sometimes respond in a very special way. At the same time the people were unwittingly acknowledging that there is a special relationship between man and dolphins.

8

Breaking the Bond

WHEN white people first started to colonize the plains of central North America, they were staggered by the vast numbers of buffalo that lived there. Similarly, when the first explorers trekked into the African continent, they were astounded by the abundance of game. The numbers of wild animals were so enormous it seemed nothing man could do would seriously affect them. We now know how wrong they were. Within a few decades millions of buffaloes were slaughtered, and gigantic herds of magnificent animals were brought to the verge of extinction. In Africa, a human population explosion, black and white, and the introduction of the gun, annihilated most of the wildlife. It is true that some of it has now been preserved in game reserves. But there are people still alive today who will tell you that the animals that remain are a thin shadow of the wildlife they saw in their youth.

For hundreds of years before these events the animals and the humans in Africa and North America were roughly in balance with one another. The North American Indians hunted the buffalo with simple weapons. Likewise the various tribes in Africa preyed upon the wild game. Man and the lion were the top predators in a balanced ecosystem. Both hunted just for food. When the gun was invented everything changed.

Approximately one fifth of the earth's surface is land; the rest is water. And what has already happened on the land is now taking place in the sea. Man is exploiting it to destruction for short-term gain. As one species is reduced to economic extinction, man turns his attention to the next. And the range of his activities extends from the largest of the sea's creatures to the smallest. The anchovy harvest off South America was the biggest in the world, producing millions of tons of fish each year. More and more vessels sailed into the area to bite ever increasing chunks out of the vast shoals, until the strain was too great. Suddenly, in the 1960s, the anchovy harvest collapsed.

A similar situation prevailed in the North Sea when the herring were overfished. But saddest of all is the decimation of the giant whales. The huge Baleen whales roamed the seas for millions of years, peacefully sweeping for plankton with their giant sieve-mouths. Now many – including the largest of them all, the Blue whale – are endangered species. Once these docile giants are reduced in numbers to a stage where they cannot breed, a miracle of creation – which took millions of years to perfect – will disappear from the face of the earth forever. And our planet will be poorer for it.

Like all life forms in the sea, the dolphins are also under threat. In the distant past, in the Pacific Ocean, a unique relationship developed between the dolphins and the tuna fish. The dolphins swam in great schools on the surface and beneath them, swimming at the same speed and in the same direction, came the tuna fish in gigantic shoals.

Fishermen have known about this relationship for decades and have taken advantage of it. Boats would set sail from places such as San Diego in California and hunt for schools of dolphins. When they were in the midst of the dolphins the men would cast their lines overboard. The dolphins ignored the weighted, baited hooks. But the tuna did not. Soon the fishermen would be hauling tuna weighing more than 100 pounds each over the side of their boat, as fast as their strength would allow.

In the 1950s the Americans changed their fishing techniques. Instead of using long lines upon which the fishermen would hook a single fish, they adopted another method based upon the purse-seine technique, which had been used for many years for catching small pelagic fishes. As a result millions of dolphins died accidentally – caught up in the tuna nets.

In 1972 the Marine Mammal Protection Act was passed in the USA. It required that fishing techniques should be modified in order to reduce the number of dolphins killed, but it is one thing to pass a law and quite another to make sure that the law is upheld, especially on the high seas, well out of sight of land. Once any captured dolphins, dead or alive, are thrown back into the sea, there is no evidence.

Conservationists had to work hard to ensure that the fishermen complied with the regulations. Even so it is accepted by those who have studied the subject that despite modifications to their fishing methods, significant numbers of dolphins will continue to die as long as large purse-seine nets are used for catching tuna fish.

In the Pacific Ocean dolphins die because, by a quirk of fate, they associate with the much-prized tuna fish. In other parts of the world the dolphins themselves are man's prey.

In the Black Sea Common dolphins, Common porpoises and Bottlenosed dolphins were hunted commercially. When gun hunting and purse-seining methods were introduced the catch increased enormously. It was estimated that at its peak, in 1938, 147,000 were caught. After that the catch declined. The percentage of mature males decreased dramatically. In the early 1960s, despite improved techniques, the kill was down to 30–40,000 animals. Most of these were pregnant and lactating females with their young. The Soviet Union introduced a ban on dolphin hunting in 1966. However, not all of the Black Sea states did the same.

Along the Izu Peninsula of Japan a different technique is used for catching dolphins. The dolphins are driven into a small inlet or fishing port. A net is then set across the mouth of the bay and the trapped animals are slaughtered for human consumption. Several different species of dolphin are caught by this method but catches, which at one time numbered 10,000 a year, are dwindling.

Porpoises are the prey of another Japanese fishery in the Iware Prefecture. About 10,000 are killed by hand harpoon between February and May, when the fishermen are not out hunting for tuna and other true fish.

Off the Japanese Island of Iki large numbers of dolphins were killed in the late 1970s – not for food – but because they ate fish. As fish populations dwindled the fishermen saw the source of their livelihood disappearing and they sought to cure the problem by annihilating dolphins. As a result, each February, when schools of migrating dolphins passed through their waters, the fishermen herded them into a bay where the dolphins were driven ashore and butchered with lances. When film of this dolphin slaughter was shown on television and reported in the press many people were horrified by what they saw and read. Widespread protests were made. Large numbers of petitions bearing thousands of signatures were sent to Japanese Embassies around the world. In 1982, the Government yielded to this pressure and asked the fishermen to stop their annual cull.

Fisherman tending tuna nets in San Diego. Countless dolphins died when the techniques of fishing were changed from hooks and lines to immense purse-seine nets.

Nothing positive will be achieved if we harbour grudges against these Japanese fishermen who were finding it harder and harder to make a living from the only method they knew how. We should express our understanding of their plight and encourage the authorities to help them retrain and find alternative employment. Such changes are always difficult to bring about because they may bring hardship and heartaches. But they have had to happen in other areas of the world – such as the port of Hull in England – where the fishing industry, which once employed thousands directly, and many more indirectly, had dwindled. As a result a once prosperous and tight-knit fishing community has now almost disappeared.

The dolphin massacre at Iki is one symptom of the disease that threatens all of the seas of the world: over fishing by man. The disease is caused by the fact that almost every country in the world is looking to the sea as a source of protein. It is difficult enough to introduce laws that will regulate the activities of a small group of countries that are geographically close together, as in the Common Market. To attempt to control fishing on a worldwide basis is at present an almost impossible task.

So harvesting the seas has become a free-for-all, with the lion's share going to those prepared to invest in massive vessels loaded with the latest technological aids. Nations that can put men on the Moon can also sweep the sea clean of its natural inhabitants, including the dolphins.

The oceans of the world came into balance over millions of years. In very recent times, in evolutionary terms, man has completely upset that balance. Only time will tell if we will irrevocably change the sea, like we have the land.

If we do so, will we break the magic bond that seems to have existed between dolphin and man since the Minoan civilization in 1500 B.C. and probably before that?

9

International Dolphin Watch

IT IS always a great moment on a boat when dolphins are sighted. Sometimes they will come from a long way off, leaping out of the water as they travel. When this happens they usually make straight for the forward end of the vessel, and weave from side to side as they are pushed along by the bow wave. Often they will jump out of the sea and put on an unforgettable display of joyful aerobatics.

Some sailors have noticed that they can attract dolphins. One skipper of a small sailing boat keeps a very large tuning fork onboard for just that purpose. If he spots a dolphin, he bangs the fork and then holds it in contact with the hull. This transmits sounds into the water which the dolphins come to investigate. Another sailing enthusiast discovered that the sound of his children singing would attract a school of dolphins when he and his family were cruising off the Riff Bank in the Moray Firth in Scotland. Sadly, however, sightings of dolphins around the coast of Britain are becoming rare.

If you went on holiday to one of the British seaside resorts in the early part of this century, it is likely that sometime during your stay you would have seen dolphins or porpoises playing in the sea. Nowadays, apart from in a few exceptional locations, you would count yourself very lucky if the same thing happened.

Why have the dolphins disappeared?

That was one of the questions I asked myself when people told me of the joy they had when watching dolphins.

In order to cast some light on the matter, in 1978 I set up a programme of research to increase our knowledge and understanding of dolphins. I called it

A Common dolphin leaps joyfully from the sea. Sights such as this have brought joy to sailors for centuries.

International Dolphin Watch. I itemised many subjects which I felt needed to be investigated; but the more information I discovered about dolphins, the more I found questions that could not be answered.

How many dolphins are there in the sea?
What do they do in the sea?
Where do they come from?
Where do they go to?
Where do they give birth to their young?

When I put these questions to one of the world's leading authorities on cetaceans, Professor Richard Harrison at Cambridge University, he said

The shape and lines of movements of dolphins have a grace and beauty which are fleeting unless captured by an artist and frozen in space and time as in this modern Greek sculpture.

simply, "Nobody knows". Quite a lot was known about the anatomy of dolphins and how they use sound to locate their prey underwater et cetera, but very little detailed information had been gathered on their day-to-day lives in the sea. I had stumbled on one of the many great mysteries of nature.

I decided to attempt to find at least some of the answers. What I needed were thousands of pairs of eyes keeping watch – and taking notes. As Conservation Officer of the British Sub-Aqua Club, I had seen the birth of The Underwater Conservation Society. Hundreds of amateur divers had become the 'underwater eyes' for a few specialist marine biologists. The information gathered was detailed and comprehensive. The scientists were delighted and the divers were enthusiastic. As individuals there was little they could hope to do to add to our knowledge of the sea; but as a group they could make a valuable contribution.

Could I apply the same principle to solving some of the mysteries of the wildlife of the dolphins? When I put the idea to Professor Harrison he said he would be delighted to co-operate. So the Dolphin Survey Project of International Dolphin Watch was born.

Working closely with Denis McBrearty, who for some time had been collecting sighting data from the Royal Air Force Air-Sea Rescue Service launches and from fishing vessels and had been associated with Professor Harrison for many years, we produced a dolphin identification chart and a Dolphin Spotters Handbook. These were the tools with which we armed our voluntary dolphin watchers. All they needed after that was the good fortune to be in the right place at the right time.

The major questions we asked them to answer were: When? Where? How many? How big? and What are they doing? We hoped that with practice they would be able to identify different species of dolphin.

The Dolphin Survey Project was a great success from the start. Most of our sightings came from those who worked at sea, such as ships' crews and men on oil rigs. Passengers on holiday cruises and members of sailing clubs also completed report forms, several people on long cruises keeping special dolphin logs in which they recorded their sightings. The dolphin log was filled-in regularly like the daily log kept by the captain of a ship. All of the data collected were recorded and analysed by computer. Thus we have started to build up a picture of dolphin populations around the world.

After I had set up International Dolphin Watch it soon became apparent to me that the dolphins, not just around Britain, but also in many other parts of the world, were diminishing in numbers. In some cases the causes could be clearly identified. I detailed some of them in my book *Save the Dolphins*. After public lectures and showings of my films in Europe and countries as far apart as South Africa, New Zealand, Australia and the United States of America I was approached by numerous people who expressed concern for the future of dolphins and who wanted to help.

As a result the International Dolphin Watch Supporters Club was set up in 1981. One of its aims was to work actively towards the conservation of dolphins, for instance, to help save the rare Indus River dolphin from

A friendly wild dolphin named Percy off the Cornish village of Portreath where Bob Holborn, who was very sensitive to the dolphin's moods, took many visitors out in his boat to experience for themselves the magic of meeting a wild dolphin in the open sea. The story is told in my book 'Tale of Two Dolphins'.

extinction. Another aim was to finance research projects. It was intended that these should be mainly of an observational nature on wild dolphins that were totally free in the sea. We proposed also to keep members of the Supporters Club informed on what was happening in the dolphin world by sending them newsletters. (For details of the International Dolphin Watch Supporters Club see p. 87.)

The latest International Dolphin Watch project is called Operation Sunflower. It is based on my belief that Nature can provide relief for many human ailments if only we can find the key. Thus aspirin occurs in the bark of the willow tree and a green mould that grows on bread and cheese contains the antibiotic penicillin which has saved countless lives. Few people will deny the fact that dolphins have an uplifting effect on the human spirit. Operation Sunflower is looking into the possibility that it may be possible to capture this special magic quality of dolphins and help people who are depressed.

Now we all have days when we are down in the dumps. Most of us get over these periods when we have the blues and bounce back into a happier state of mind. But there are people who just cannot do this and their symptoms can take different forms. Most withdraw into themselves, they become sullen and do not communicate. They are often highly intelligent sensitive people for whom the pressures generated in our modern materialistic world are just too much to bear. It is estimated that in years to come one person in ten will need some psychiatric help during their lifetime.

Dolphins have not created possessions and they do not need houses to live in. There is no money in the dolphin's world and no basis for greed or envy. The pressures of keeping up with the Jones's therefore does not exist. The dolphins always appear to be happy with their lot in life regardless of their circumstances. And it may be this quality that they pass on to humans who come into contact with them. Whatever it is, watching them and being with the dolphins makes us humans feel good inside.

I talked about this quality of dolphins during a film show I gave in 1986 in the little fishing village of Solva in Wales. I was there making a film entitled *Bewitched by a Dolphin* for Harlech Television. After my presentation I was asked if I would take a man out to see the wild dolphin called Simo we had been filming. His name was Bill Bowell and I was told he had been unable to work for twelve years because he was suffering from chronic depression.

He had been given many forms of treatment but modern medicine could not help. Could Simo the dolphin? Taking Bill out to see the dolphin certainly would not do any harm. So I agreed. And the next day Bill, who was fifty two years old, put on one of my old wetsuits and went into the sea with Simo. That meeting changed his life. When Tricia Kirkman, who was working with me on the film about Simo, commented, "Bill has blossomed like a Sunflower," I decided to investigate the effects dolphins might have on depressives. And so the project we now call Operation Sunflower was born.

The first major step forward in the project was made in 1987 when I was able to take three people with different types of depression to Dingle in Ireland to meet a friendly wild dolphin who was sometimes called Dorad and sometimes Fungi. One of those who met the dolphin was a brilliant 20 year-old student called Jemima who suffered from anorexia nervosa. She was very pretty but weighed only 41 kilogrammes (6½ stone) because she could just not make herself eat a normal diet. Neal was our second subject. He was 24 years old and had a job in a laboratory but sometimes his body froze and he could not get out of bed in the morning or do up his shoelaces because he felt people were getting at him. The third person was Bill who had already met Simo the dolphin off Solva.

We filmed the effect the dolphin had on our three patients, and the effect they had on the dolphin. A year later we all returned to Dingle to complete a film entitled *The Dolphin's Touch* for the TVS series *The Human Factor*.

It would be totally impossible to take out to meet wild dolphins all of the people who need psychiatric help. So the next stage of the project is to see if we can take the dolphins into psychiatric wards. Not physically of course, but via video recordings. Music can affect the way we feel and we do not need a composer or a musician within 100 kilometres for the experience. We can capture the magic mood-changing power of music on disc or tape, re-create it and even enhance it with the aid of modern micro-chip technology. Only time will tell if a similar thing can be done with the magic uplifting power of the dolphins. Whatever the outcome the full story of Operation Sunflower will be told in a new book entitled *Dance to a Dolphin's Song*.

10

Making Close Contact

OPERATION SUNFLOWER would have been impossible if the dolphin in Dingle had not been one of those rare dolphins who seek out human company in preference to that of his own kind. As news of his friendliness spread, the number of visitors to the fishing town in County Kerry in southwest Ireland increased. During the summer of 1988 some of the locals abandoned inshore fishing in favour of taking tourists out to see the dolphin. One of the visitors, a nun, sister Emanuel Immaculata from Limerick, who was over seventy years old, put on a swimming costume and swam with the dolphin at the edge of the Atlantic Ocean. Although the water was chilly and she did not have the protection of a wetsuit, when the nun was hauled back into the fishing boat she said she had not felt the cold because she was so excited at meeting and touching the dolphin. She was benefitting from a dolphin–human relationship that had been building up over several years.

When Donald arrived off the Isle of Man in 1972 those who befriended the dolphin were breaking new ground. His arrival coincided with the time when interest in the relatively new sport of diving was increasing rapidly. By insulating themselves against the cold with wetsuits, humans were able to spend hours in the sea off the British coast. By using aqualungs they were even able to breath underwater. Thus for the first time in the relatively cold waters around northern Europe humans were able to swim freely and comfortably in the dolphin's environment. When some of them discovered the joy and excitement of cavorting with Donald in the wide open sea it became many a trainee diver's dream to swim with a dolphin. When Donald disappeared in 1978, lots of sub-aqua enthusiasts were left with their vision unfulfilled.

Percy, a friendly wild dolphin off the Cornish coast, s.w. England.

So, when Percy arrived off Portreath in 1982 there were plenty of humans who were not only willing and able, but very keen to respond to his advances. News of his extraordinary antics spread rapidly, mainly as the result of television features. Many divers made their way to Cornwall to swim and play with him.

News of the fun of interacting with humans now seems to be spreading throughout the dolphin world – in Europe at least. In the December 1988 edition of the International Dolphin Watch Newsletter, *Dolphin*, details were given of no less than seven friendly solitary dolphins all seeking out human

Regular boat trips set out from Sheppard's Marina in Gibraltar to take visitors on dolphin safaris. Passengers can lie on the deck and touch the dolphins that come to play between the hulls of the catamaran. The skipper, Mike Lawrence, has a special love and knowledge of dolphins.

company. Their locations ranged from Ireland in the west to Yugoslavia in the east. One was a young Killer whale who during the summer had been befriending men working on an oil rig in the North Sea. The others were all Bottlenose dolphins in more easily accessible locations. For those living in England the most exciting news was that of a dolphin swimming in an apparently small and well defined territory off the mouth of the river Amble in Northumberland. He had been dubbed Dougal by the local divers who regularly went out to swim with him, and they reported he was extremely friendly. It was impossible to predict just how long he would stay in the

73

Interaction with dolphins can only occur if the dolphins themselves choose to participate.

vicinity but everyone hoped he would become a long-term resident like, Jean Louis who had remained in the same location in Brittany for ten years.

1988 was a good year in Europe for those seeking a close encounter with a friendly, wild, solitary dolphin. However, dolphins normally associate in groups we call pods or schools. So what are the chances of making close contact with a family of dolphins?

Dolphins can surface and breathe leaving barely a ripple. Once underwater they can disappear out of visible range in a second or two. They can stay submerged for several minutes and in that time can travel long distances because of the speed and efficiency with which they can swim. Thus, if dolphins decide they do not want to be detected they can apparently just vanish from the scene. This makes studying and interacting with them in their

74

A human-dolphin encounter in the sea is always a memorable experience.

75

natural environment extremely difficult, and it can only be done if the dolphins themselves choose to participate. So if we want to watch a group of dolphins we have to go where they are likely to appear and hope that they will divert from whatever activities they are engaged in and come and say "Hello".

In the past underwater encounters between divers and schools of dolphins have taken place on an apparently random basis. In his book *Dolphin Dolphin*, Wade Doak has described the exciting events that took place when he and his wife Jan met dolphins off the coast of New Zealand and then made contact with the same group at later dates.

In 1976 a group called *Friends of the Sea* (address is given on page 87) was set up in the USA. One of its aims was to find out more about dolphins in their natural habitat. To do this they paid visits to a group of Spotted dolphins that had earlier become friendly towards a group of treasure-hunting divers working on Little Bahama Bank which is about 80 kilometres (50 miles) off the coast of Florida. Scientists, musicians and others simply with a love of dolphins joined Friends of the Sea and tried a wide variety of methods of interesting and interacting with the dolphins. The results, recorded on film and video, provided a visual record of the continuing build-up of a joyous relationship between a community of dolphins and many different types of humans. The organization now operates week-long boat trips on a limited basis, in which paying guests are able to meet the dolphins and frolic with them in warm, clear-blue water in the wide open sea.

As trust builds up and the news spreads in both human and dolphin populations this approach is providing a model for enterprises in which scientists and holidaymakers can come together in an informal atmosphere onboard a boat to study and simply enjoy the company of free-roaming wild dolphins.

All of this is fine if you are a swimmer and enjoy sailing. But what if you can't swim? Well there is one place in the world where an entire group of dolphins have chosen to come to the beach and will actually swim between your legs if you just paddle in the sea. That place is called Monkey Mia and it is about 800 kilometres (500 miles) north of Perth in Western Australia in a region called Shark Bay. It is a remote place in a wilderness area which means that those who go there have to make a big effort to do so. But for dolphin devotees it is well worth the time and effort needed to make the trip.

Monkey Mia in Australia is the only place in the world where wild dolphins regularly come inshore to be stroked.

I was lucky to be able to visit Monkey Mia in 1986 and see and experience for myself the special magic of the location and the dolphins.

I went as the guest of Wilf and Haze Mason, the proprietors of the Monkey Mia Caravan Park, which is adjacent to the beach where dolphins come to the shore. I was fortunate enough to have their young granddaughter, Rebecca, as a guide. She knew the names of all of the dolphins and introduced me to them.

Many people think dolphins are all similar, like fish. But they are not. They have different personalities and with experience it is possible to identify each dolphin by its dorsal fin which is often chipped and scarred. The one dolphin that nobody can mistake at Monkey Mia has a hole in her fin and is called Holey Fin.

The author and Rebecca feeding Holey Fin and Holly at Monkey Mia.

Most people arrive at Monkey Mia by car or coach. The nearest resort is Denham. I drove from Perth to Denham and made the final part of my journey after a rare rainstorm. The road, which was unsurfaced, was a quagmire in places and I got myself plastered in red mud when I helped a group recover their vehicle which had skidded and turned over.

I was lucky when I eventually arrived at the campsite because some of the dolphins were 'in'. They were patrolling up and down the beach and I was able to go paddling with them immediately. I say I was lucky because the dolphins are certainly not always present and sometimes, although it is very rare now, they may not come to the beach all day.

Small buckets of fish for feeding the dolphins could be purchased from the warden.

79

Dolphins moving into the bay at Monkey Mia in the early morning.

Prior to my visit to Monkey Mia I had read Elizabeth Gawain's book entitled *The Dolphin's Gift* and had corresponded with Wilf and Haze Mason. When I eventually met them I found the Masons to be an extremely homely couple. After our introductions it wasn't long before Wilf, wearing a wide-brimmed hat and sitting astride a little motor scooter, was escorting me to the caravan which was to be my home for the next four days.

I knew the Masons had a very deep-rooted and sincere concern for the well-being of the dolphins. It was largely through their efforts that the 'Dolphin Welfare Foundation' was first set up to inform visitors of the very special relationship man has with dolphins, as well as covering educational aspects of dolphin biology. As a result visitors to Monkey Mia are better prepared to

80

appreciate the experience as and when they come into close contact with the free dolphins.

A full-time warden had been appointed to patrol the beach and keep a watchful eye on the dolphins and the humans. The warden operated from a purpose-built centre, erected very close to the water's edge. Here information sheets were always available. At certain times of the day visitors could buy fish to feed to the dolphins. But it wasn't only the dolphins who benefitted from these handouts. There were several pelicans who would pinch fish from the buckets of unsuspecting dolphin feeders who put their buckets down unattended on the beach for a few moments. I even saw one very cheeky pelican snatch from a person's fingers a fish just as it was about to be dropped into the open mouth of an expectant dolphin.

I was able to share briefly the Monkey Mia experience with millions of viewers during the launch of my book, *Tale of Two Dolphins,* when some film I shot at Monkey Mia was shown on the BBC TV programme *Blue Peter.*

It was the dawn of each day at Monkey Mia that was especially magic for me. At the first sign of light I would get up, put on a sweater and shorts and make my way down to the beach. I would scan the flat grey sea for the dolphins. Invariably I saw no sign of them. So I would kick off my shoes and wade into the water. I loved the solitude of those moments, listening to the gentle murmur of the sea as it swished up the sandy beach. A rose glow would tint the sky and then spread upwards like a fire coming alight. I would stroll back and forth in the sea, stopping occasionally to dig my toes into the sand until my feet felt as if they were being sucked down into the ocean floor – at the interface of two worlds – mine and that of the dolphins.

Then, as if from nowhere, they would appear. My first sighting was usually a few metres away when I would see a grey curved fin appear briefly and silently and heading towards me. It was usually Holey Fin. A few seconds later her domed head would break the surface in front of me. With jaws open she would raise her head in greeting. After saying "Hello" I would shape my hands into a cup, scoop up some water and let it dribble on to her.

She seldom came alone. A few moments later her daughter, Holly, who was born at Christmas-time, and perhaps Crooked Fin would join her. I would have the company of three dolphins gliding smoothly around me – a dawn reverie of dolphins – during which the rim of the poppy-red sun would

Holidaymakers enjoying the unique experience of feeding and playing with dolphins.

slowly and silently emerge on the horizon and pour its light on to the sea.

Then the dolphins would move silently away swimming very close to the shore. They were going to greet another human who had just walked into the sea. He was a lad about sixteen years old who was very sensitive and gentle. I knew he wanted to spend a few moments alone with the dolphins before we waded towards each other to exchange words.

As the sun rose, so too, did most of the occupants of the camp site. Some took a pre-breakfast stroll along the beach. Others paddled into the sea and stroked the patrolling dolphins. Campers clutching toilet bags and towels made their way to the central shower facilities – as campers do everywhere in the world.

Is Holey Fin conducting her own research on this scientist from an American university?

About one hour after getting up I would usually return to my caravan to make a cup of tea whilst my soul floated on its own sea of peace and tranquillity. After breakfast the atmosphere would have changed completely. Boats would be launched to set off on fishing trips. Families would engage in the time-honoured pursuits of quiet seaside holidays: building sandcastles, strolling, sunbathing and swimming.

However for the lucky holidaymakers at Monkey Mia there was also the possibility that they might be able to do something that could not be done anywhere else in the world. To feed, stroke and play with a group of dolphins who were totally free to swim off at any time yet who came to shore to share their friendship with humans.

The dolphins at Monkey Mia will say 'Hello' even if they are not offered fish.

Nobody is sure why the dolphins should choose to meet humans at Monkey Mia. It has been suggested that it was the fishermen who first attracted the dolphins to the shore by tossing them spare fish from the wooden jetty. If the explanation was as simple as that then it would not be unreasonable to assume that similar situations would have occurred elsewhere. But they have not. So the mystery remains.

As we have already seen there is no way we can impose our company on dolphins if they do not wish it. So we must accept the fact that at Monkey Mia it is the dolphins, not the humans, who have chosen to maintain contact. This being so it is tempting to speculate why families of dolphins should repeatedly come to the shore to exchange greetings with families of humans. Is this a deliberate attempt on the part of a nomadic tribe of intelligent sea-beings to

A young visitor exchanging a kiss with a dolphin at Monkey Mia.

make contact and establish friendship with what to them would be an alien land-based intelligence, i.e. us humans? Support for this seemingly unlikely suggestion comes from the fact that in addition to offering friendship, the dolphins are now sometimes bringing gifts of fish from the sea and tossing them to people on the beach.

There are millions of dolphins throughout the world and hundreds of places where the dolphins could have built-up close, friendly contact with humans. So why did they choose Monkey Mia?

Was it because it is in one of the few wilderness areas left on earth where humans have not sought to subdue the wildlife and bring the environment under control in order to exploit it?

How long has contact really been made?

Was the connection made a long time ago by the dolphins with the Aboriginal Australians who had a profound love, knowledge, understanding and above all respect for all forms of life in their territories? Was that contact broken earlier in the century? If that was so, did the White Australians who tossed fish to the dolphins just happen to remake that contact by chance?

Whatever the answers to these questions, the fact remains that a metaphorical bridge has been built at Monkey Mia. It is a bridge over which humans can pass to sample the magic world of the dolphins. It is a bridge that all those with a concern for the future well-being of our planet should cherish and strenuously seek to preserve. Those lucky ones who have crossed over that bridge will appreciate the following words written in his poem Halieutica by Oppian in the 2nd Century A.D.

> *Diviner than the dolphin is nothing yet*
> *created for indeed they were aforetime*
> *men and lived in cities along with mortals,*
> *but by the devising of Dionysus*
> *they exchanged the land for sea and*
> *put on the form of fishes.*

Useful Addresses

International Dolphin Watch, Parklands, North Ferriby, Humberside HU14 3ET, England.
Friends of the Sea, P.O. Box 2190, Enfield, Conn. 06082, USA. Telephone (203) 749-0256
Dolphin Watch, P.O. Box 4821, Key West, Fla. 3304, USA. Telephone (305)294-6306. Dolphin Safaris led by Captain Richard Canning.

Acknowledgements

I express my thanks to all those who have helped to make this book possible especially my friends who have kindly provided photographs. These include Nigel Rolstone – otherwise known as 'Animal' – who ranks his encounter with Percy, the wild dolphin, off the Cornish coast amongst his most memorable experiences; Georgie Douwma and Peter Scoonies with whom I spent a memorable couple of weeks studying and filming the wild dolphin Jean-Louis off the French coast; Phyllida Cotton – whose father Dr. Simon Cotton kindly took me out searching for dolphins and killer whales off the coast of New Zealand; Penelope Otto whom I also met in New Zealand and who gave me her photograph of the Opo statue; Denis Moor, whose delightful company I enjoyed in Crete; Chris McLoughlin who was the perfect host in my search for Sandy – a wild dolphin in the Bahamas; Angela Ames who took the picture of the dolphin skeleton found in the Galapagos Islands; Arnold Madgwick at the Institute of Oceanographic Sciences; David Slinger who spent many hours with the dolphin school that frequents Little Bahama Bank; and my filming partner Chris Goosen without whose enterprise I would never have found Dobbie the dolphin in the Red Sea, or heard Moses Benga's story in Sierra Leone. My secretary Kerry Davis who transcribes my tapes and untidy hieroglyphics into impeccably typed manuscripts is warmly thanked for her efforts, as is Doug Goodwin for his work on my pictures in the darkroom. To Denis McBrearty, who kindly reviewed the manuscript of this book, and Gillian King, both at Cambridge University, I send my sincere appreciation for sharing with me their expertise and knowledge of dolphins. I thank Dr. Robbins Barstow and his friends in the Connecticut Cetacean Society for their hospitality during my visit, in particular Bill Rossiter for permission to use his pictures of Killer whales in this book. I thank Wilf and Haze Mason for inviting me to Monkey Mia and for the hospitality they gave me during my visit.

Credits for Photographs

Nigel Rolstone 68; Chris McLoughlin 11; Phyllida Cotton 13; Angela Ames 19; Denis Moor 26; Bill Rossiter 31 and 34; George Douwma 46; Chris Goosen 51; Penelope Otto 55; Arnold Madgwick 65; David Slinger 74. All other photographs including the jacket cover were taken by the author.

About the Author

Horace Dobbs, B.Sc., Ph.D., founded The Oxford Underwater Research Group in 1963, and is also founder and director of International Dolphin Watch – a non-profit making organisation for the study and conservation of dolphins.

A Fellow of the Royal Society of Medicine, he gave up a successful career in medical and veterinary research in order to devote himself full-time to his passion for dolphins and underwater exploration. He now earns his living as a writer, freelance broadcaster and film maker, and also lectures widely. In 1978 he was awarded a Silver Bowl by the International Platform Association in Washington, U.S.A., for his outstanding film/lecture presentation on dolphins, and the following year gave a major film presentation on dolphins at the National Geographic Society, also in Washington.

Dr Dobbs was one of Britain's pioneers in underwater photography and film making, and he has led diving expeditions to many parts of the world. His first book, *Camera Underwater*, became the standard work on underwater photography. In 1985 he was awarded a gold medal by the Chairman of the British Sub-Aqua Club for his outstanding services to diving.

His full-length underwater film, *Neptune's Needle*, which was shot exclusively in British waters, won a gold medal at the International Film Festival in Brighton in 1966. His latest film, *The Dolphin's Touch* is about Operation Sunflower.

In 1986 Horace Dobbs was awarded a gold medal at the Athens International Festival of Underwater Films and Photography for his book *The Magic of Dolphins*.

Other Books by Horace Dobbs

Camera Underwater (Focal Press, 1962)
Classic Dives of the World (The Oxford Illustrated Press, 1987)
Dolphin Spotters Handbook (International Dolphin Watch (with R. J. Harrison, D. A. McBrearty and E. Orr) 1979)
⋆*Follow A Wild Dolphin* (Souvenir Press, 1977)
The Great Diving Adventure (The Oxford Illustrated Press, 1986)
⋆*Save the Dolphins* (Souvenir Press, 1981)
Snorkelling and Skindiving (Oxford Illustrated Press, 1976)
Tale of Two Dolphins (Jonathan Cape, 1987)
Underwater Swimming (Collins, 1966)

⋆ A combined version of these books entitled *Follow the Wild Dolphin* is published in the U.S.A. (St Martin's Press, 1982)